In search of the . . .

ORIGIN OF LIFE

EVOLUTION • CREATION

RICHARD B. BLISS, Ed.D.

Director of Curriculum Development
Institute for Creation Research

GARY E. PARKER, Ed.D.

Head of Natural Sciences
Clearwater Christian College

DUANE T. GISH, Ph.D.

Vice President
Institute for Creation Research

C·L·P PUBLISHERS
San Diego, California

ORIGIN of Life

First Edition   1979
   Third Printing   1990

Copyright © 1979
C·L·P· PUBLISHERS
P.O. Box 1606
El Cajon, CA 92021

Library of Congress Catalog Card Number 78-58477
ISBN 0-89051-053-9

Cataloging in Publication Data
Bliss, Richard Burt, 1923-
   Fossils: key to the present
   1. Paleontology.   I. Parker, Gary E. 1940-
II. Gish, Duane T. 1921   III. Title
                                577   78-58477
ISBN 0-89051-053-9

Printed in the United States of America

ABOUT THE AUTHORS

Richard B. Bliss, Ed.D., has more than 39 years experience in all areas of science education. In Addition to having taught chemistry, physics, biology, and general science at the high school level, he was adjunct professor, teaching science methods to teachers in the University of Wisconsin System. He was engaged in biological research and obtained several National Science Foundation grants and fellowships during his academic career. Recently he developed a hands-on curriculum in science for K-6 elementary students that is based on the most current research in science education. Dr. Bliss is a frequent speaker on the creation/evolution issue in the U.S. and other countries. He is presently professor of science education and Chairman of the Science Education Department at the Institute for Creation Research in Santee, California.

Gary E. Parker, Ed.D., formerly ICR Research Associate and Professor of Biology, is now Head of Natural Sciences at Clearwater Christian College, Clearwater, Florida. He did his doctoral work in biology (amphibian endocrinology) and geology, and is the author of five widely used program instruction books in biology, including *Life's Basis: Biomolecules* (Wiley), and *DNA: The Key to Life* (Educational Methods). Dr. Parker, a member of Phi Beta Kappa, has received several competitive scholarships, including a National Fellowship Award from the National Science Foundation.

Duane T. Gish, Ph.D., is Vice President of the Institute for Creation Research in California. He spent 18 years in biochemical research at Cornell University Medical College, the Virus Laboratory of the University of California, Berkeley, and The Upjohn Company, Kalamazoo, Michigan. He is the author and co-author of numerous technical articles in his field and is a well-known author and lecturer on creation/evolution. He received his Ph.D. in biochemistry from the University of California, Berkeley.

ACKNOWLEDGMENTS

We wish to express our deep appreciation to the many science teachers, administrators, and students whose valuable reviews have been a great encouragement as we have offered the student this open and scientific approach in the *Origin of Life*.

Richard Bliss

TO THE STUDENT

The subject you are about to study in this module is the *origin of life*. This topic has been pondered by scientists over the centuries and many believe that they have evidence to support their views, yet no one was there in the beginning to see what really happened. Laboratory experiments have been conducted to see if these first elements and hypothesized compounds could ever have produced even the smallest evidence of living substance, and governments have spent billions of dollars in search of life on other planets.

On the other hand, there are scientists who are analyzing all of these data from a special creation point of view. They are asking pointed questions which relate to the intricate design found in living substances and in man. The arguments you will encounter are clearly scientific, and must be considered with an open mind.

You will find that this module doesn't attempt to balance the data to make one side equal to the other, in space. However, every attempt has been made to bring forth scientific criticism to bear on the subject. This has been done in order to call your attention to the importance of a critical examination of the data. To help you, we are following our investigator, as he goes about examining the basis for various assumptions. The data are clearly stated, and scientists' views that are significant are included. You be the judge! Do the facts fit the *creation* model better than they fit the *evolution* model?

CONTENTS

CHAPTER ONE

Two Models for Scientific Investigation

Perhaps you have read the stories or watched the TV news reports about the search for life on Mars. When you see the pictures of the barren craters on the moon or the rocky emptiness of Mars, maybe you have wondered why our planet Earth has been blessed with such an abundance of living things.

What were conditions like on Earth when life began? No one really knows. In fact, billions of dollars are being spent to satisfy man's inquiring mind concerning the origin of life.

The majority of scientists today think that life originated by natural processes of *chemical evolution*, but a growing number of scientists think that all life was brought into being by supernatural acts of *special creation*. So, in our study, we will approach the origin of life from a *"two-model"* point of view. In the best interests of science and science education, we will be asking you to consider the question: Do the facts fit the *creation* model better than they fit the *evolution* model?

The answer can only come from your own critical analysis of the data. Yes, *you* will have to finally decide for yourself on this all important question. There is little value to you in having your teacher or anyone else make this decision for you. Your reasoning will be most important, and your views on the subject may not agree with others.

When a person looks at the origin of life from a "two-model" approach, he or she finds an exciting challenge. The whole idea involves sophisticated *detective work*; but even more challenging is the opportunity to examine some of the exciting *data* now in hand.

This module will involve much of the current information, principles, arguments, and experiments that can be brought to bear on the origin of life. You will be asked to examine the data and give your opinion based upon the evidence available. Do these data

seem to show that evolution could or could not have occurred? Do the data seem to show that a supernatural designer was or was not involved in creating the first living organisms?

The Scientific Approach to the Evolution and Creation Models

The *creation* model, like the *evolution* model, is subject to a degree of experimentation. It can be used to explain observations and to make logical predictions concerning new discoveries within the rules of science.

How can we gather scientific data in favor of a *creation* model? The most natural and reasonable approach is to determine how an object created (designed) by man could be identified apart from an object produced by natural processes. Consider a television set, a landscape painting, and an automobile as examples. We know enough about copper wire, oil paint, and gears to know that these objects could not make themselves, no matter what amount of time was involved. Scientists can likewise look for evidence of *creation* in the kind of design found in living things.

But not all examples of apparent design or complexity point to a designer. Snow crystals, for example, seem to have a beautiful and precise design, yet scientific studies show that they are the result of water molecules "doing what comes naturally" according to their inherent properties. Evolutionists believe that living things, like ice crystals, have only the kind of design that results from time, chance, and the natural properties of the molecules involved.

So, when we compare the *evolution* and *creation* models, we must always be trying to answer the question: Is the design determined from the *"inside,"* (by time, chance,and natural processes), or is it determined from the *"outside"* (by a creative agent putting molecules together according to a plan or purpose beyond the ability of the molecules themselves)?

Figure 1.1 *A water molecule consists of two hydrogen atoms joined to one oxygen atom.*

Let's see how good a detective you are when it comes to examining evidence for *evolution* or for *creation*. Below are two rock formations (Figure 1.2), each made of alternating layers of hard and soft rock and each with the same appearance of design (looking somewhat like a man's head, like some formations in the southwestern deserts).

1. Which formation seems to be the logical result of time, chance, and the natural processes of weathering and erosion?

2. Which formation seems to be the logical result of creative design, for example, of an Indian craftsman shaping the hard and soft rock into a pattern unlike what natural process alone could produce?

3. If you found that molecules in living things were arranged in concept like the hard and soft rock in formation A, would you conclude that living things were created (by some special creative agent) or that they evolved (by molecules doing what comes naturally)?

4. Does the appearance of design itself provide evidence of creation, or can some kinds of design result from time, chance, and natural processes?

5. Do you have to see a creator to find evidence of creation, or can you find evidence of creation by studying relationships observed in created objects?

Figure 1.2 *Both formations look like a man's head, but one pattern of hard and soft rock was cut by weathering and erosion and the other was carved by an Indian craftsman. Which pattern must have been purposefully created?*

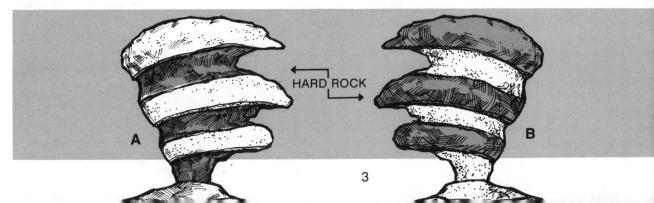

HARD ROCK

A

B

3

The Structure of Living Things

We are now ready to examine living things the same way you just examined the rock formation on the preceding page. Do we see within these living systems the kind of design that could only have been caused by an external architect or, on the other hand, do we see the kind of relationships that result from time, chance, and natural chemical processes? Let's prepare to investigate these questions about life's origin by first reviewing what living things are made of.

Evolutionists and creationists agree that living systems are made of "dust from the ground." That is, the chemical elements in living things are the same elements we find in rocks and other non-living things. In fact, living things consist mostly of just six kinds of *atoms* or chemical elements: carbon (C), hydrogen (H), oxygen (O), nitrogen (N), phosphorus (P), and sulfur (S). These atoms are found combined to form such common *molecules* as water (H_2O), carbon dioxide (CO_2), methane (CH_4), and ammonia (NH_3).

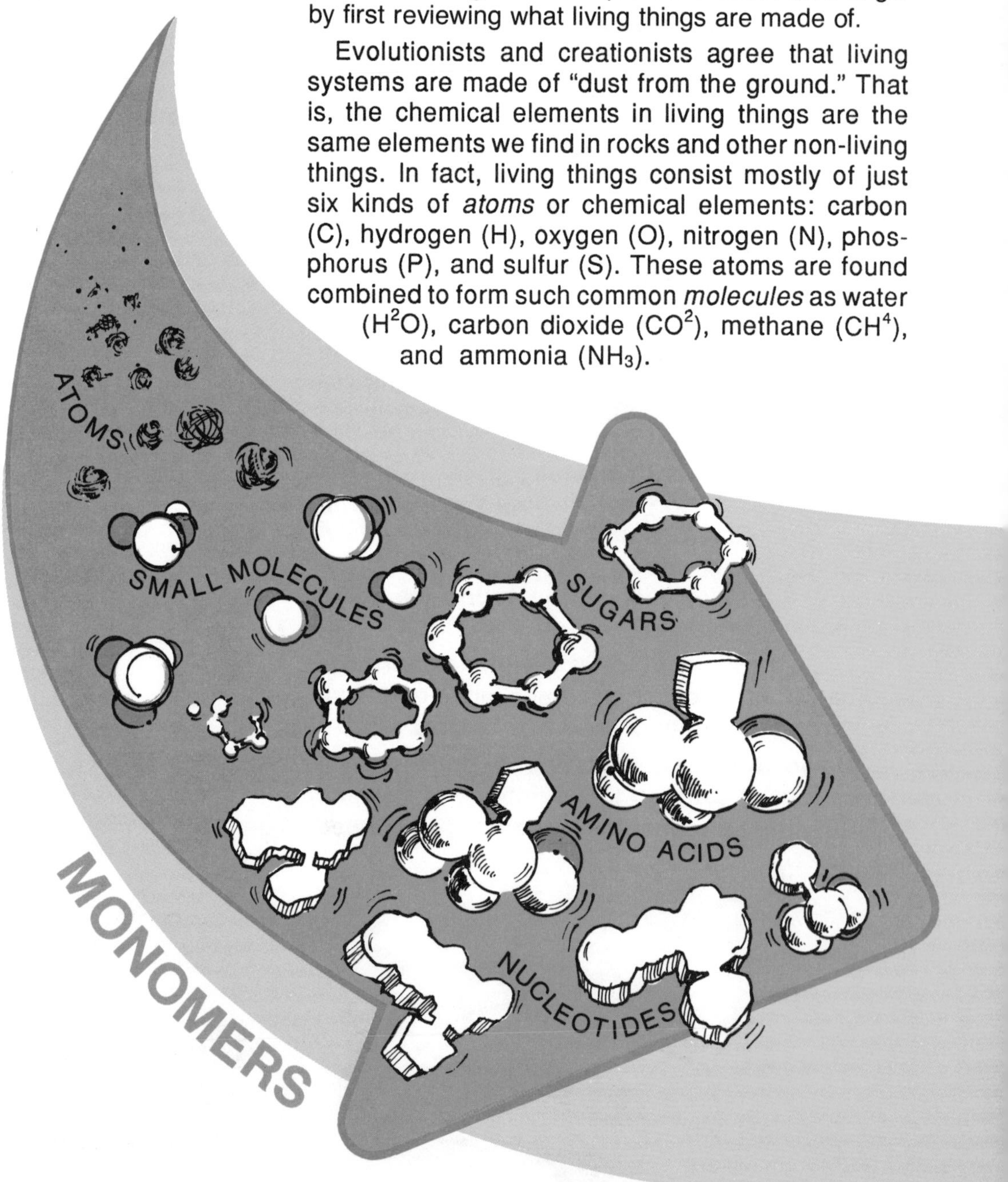

ATOMS

SMALL MOLECULES

SUGARS

AMINO ACIDS

NUCLEOTIDES

MONOMERS

In living things we also find special large molecules called **polymers**. Each *polymer* is a chain of repeating units somewhat like a pearl necklace or string of beads. Three important *polymers* and the beads, or **monomers**, that make them up are:

MONOMERS		POLYMERS
SUGARS ⟶		STARCHES
AMINO ACIDS ⟶		PROTEINS
BASES (NUCLEOTIDES) ⟶		RNA/DNA

The smallest unit of life, the *cell*, consists of millions of these molecules. But none of the molecules in a living cell is alive by itself. A dead body, after all, has the same molecules the living body had only moments before. So what makes the cell alive? *Organization*. Flying isn't a property of metal, for example, but metal and other parts organized into the shape of an airplane can fly. In a similar way, non-living molecules are organized to form a living cell.

polymers: large molecules made up of smaller molecular units (monomers) joined together in long chains

monomers: small molecular units of the same type that join together in long chains (polymers)

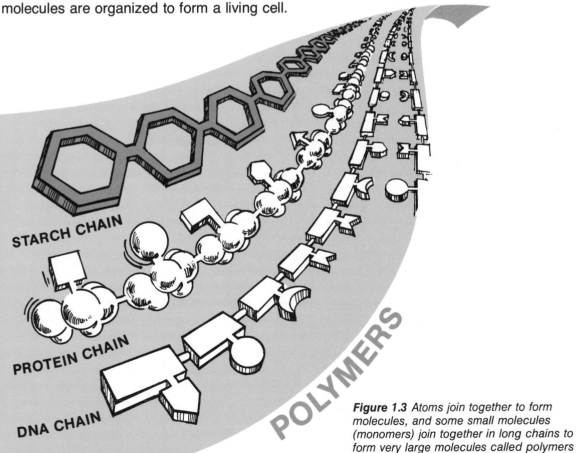

STARCH CHAIN

PROTEIN CHAIN

DNA CHAIN

POLYMERS

Figure 1.3 *Atoms join together to form molecules, and some small molecules (monomers) join together in long chains to form very large molecules called polymers (like starches, proteins, and DNA).*

ORIGIN OF LIFE

ASSUMPTIONS

DATA

How do lifeless molecules become organized into living cells? (1) According to the *creation model*, this ordering process requires and reflects the special acts of an intelligent Creator, somewhat like the manner in which the arrangement of the parts of a watch reflect the work of the watchmaker. (2) According to the *evolution model*, this organization gradually came into being over millions of years, as a result of time, chance, and molecules doing what comes naturally, because of their chemical properties.

Now we're ready to investigate the origin of life. Get out your note pad and sharpen your pencil. Be careful to sort out **data** from **assumptions** or **questions**, and keep examining your own **assumptions** and how well they help you understand the **data**. Let's bring all-out logic and scientific wisdom to bear on the subject, and ask ourselves the question: Do the facts fit the *creation model* better than they fit the *evolution model*?

data: factual information used as a basis for reasoning

assumptions: ideas taken for granted or accepted as proof, and usually used as a basis for reasoning

HOW DO YOU CHOOSE THE BETTER MODEL?

OBJECTIVES

When you finish studying this module you should be able to:

1. **Describe how scientists look for evidence of creation and explain the evidence they find.**

2. **Describe and explain the stages in life's origin as an evolutionist sees them.**

3. **Give the** *data* **used to argue for and against either evolution or creation, and explain which view you think is stronger, and why.**

4. **List the basic** *assumptions* **of the evolution and creation models, and compare them with** *assumptions you* **are willing to make about life's origin.**

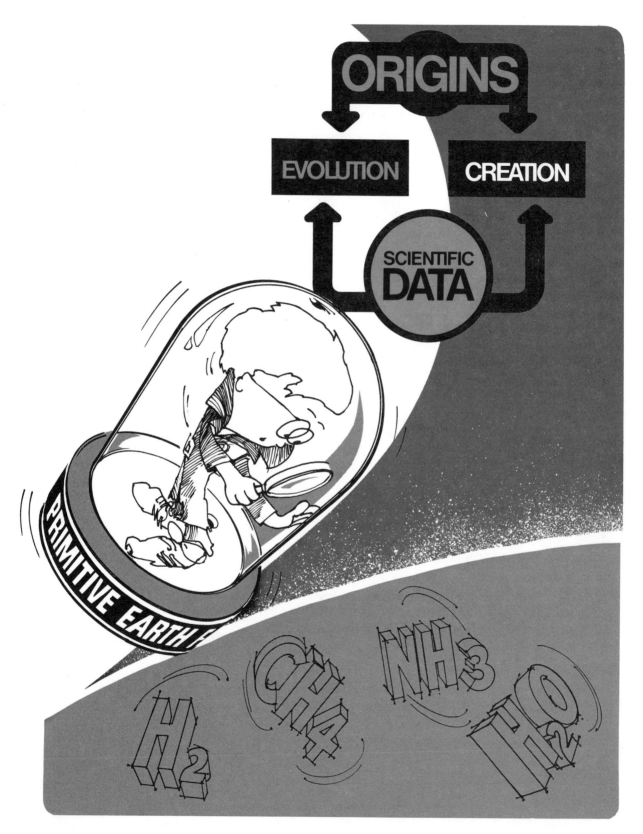

CHAPTER TWO

Evolution Model for the Origin of Life

According to the *evolution* model, the story of life's origin begins nearly 5 billion years ago, and gradually unfolds through a series of five stages.

In *Stage I* we find the "early Earth" with quite a different atmosphere than it has now. Our present **oxidizing** atmosphere, which is 21% free oxygen (O_2) and 78% nitrogen (N_2), contrasts with the idea that the early Earth had a **reducing** atmosphere consisting mostly of methane (CH_4), ammonia (NH_3), hydrogen (H_2), and water vapor (H_2O).

oxidizing: a chemical environment rich in electron acceptors like oxygen

reducing: a chemical environment rich in electron donors like hydrogen

As ultraviolet radiation, electric discharge, and high-energy particles bombard molecules in this atmosphere, we move to *Stage II*, which is the formation of small organic molecules such as *sugars, amino acids*, and **nucleotides.**

nucleotides: building blocks (monomers) of DNA or RNA, each one made up of a phosphate group, a sugar, and one of four bases (G, C, A, or T in DNA; G, C, A, or U in RNA)

starches: a mixture of polymers made of the sugar glucose

If all this happened billions of years ago under a *reducing* atmosphere, then at *Stage III* we have a combination of various small Stage II molecules forming large polymers, namely, the **starches, proteins,** and **nucleic acids** such as **DNA**.

proteins: long chains (polymers) of amino acids coiled and folded into special shapes

nucleic acids: DNA and RNA; long chains (polymers) of nucleotides each with a special base sequence (e.g., GGACATCTT . . .)

DNA: deoxyribonucleic acid, the "supermolecule" that stores hereditary information; two long chains (polymers) of nucleotides base-paired together and coiled as a double helix

At *Stage IV* these large molecules join together as gel-like globs called **coacervates** or **microspheres.** These *coacervates* could also attract smaller molecules to themselves, forming structures that might be called *proto-cells*.

RNA: ribonucleic acid, the long-chain molecule that encodes the information in DNA for protein synthesis, and which also performs other functions. In DNA, the sugar is deoxyribose; in RNA, it is rebose

coacervates: combinations of two complex compounds such as proteins and fats, which form tiny gel-like globs in water solution

microspheres: unorganized droplets of miscellaneous protein-like molecules

Finally, at *Stage V* at least one of these globs absorbs the right molecules for self-reproduction and we have a living cell. These early cells at first fed on the molecules left over from earlier stages. But soon **photosynthetic** cells evolved, and these cells released into the atmosphere the oxygen needed for

photosynthetic: process in green plants that produces sugars and oxygen from CO_2 and H_2O using light as the source of energy

most forms of life on Earth today. This oxygen as well as the feeding habits of the early cells destroyed all the molecules from earlier stages. After life evolved once, it could not evolve again.

The story of chemical evolution through these five stages certainly sounds reasonable, interesting, and even exciting. So the student often accepts the *assumptions* and conclusions presented without really looking closely at the evidence.

Are there any major roadblocks on the evolutionary pathway to life? Remember that no witnesses were there to see what happened, so our evidence is only *circumstantial*. Does the web of circumstantial evidence establish the evolutionary case "beyond a reasonable doubt," or must we look for other interpretations? Examine the *data* related to each of the five stages, and see if you can decide.

Stage I

The First Atmosphere

Ideas about the first atmosphere that existed on planet Earth immediately generate a division among scientists. There are those that insist that the first gases in this atmosphere were *reducing* such as methane (CH_4), ammonia (NH_3), water vapor (H_2O), and hydrogen (H_2).

This idea was suggested first because scientists knew that free oxygen (O_2) like we have in our *oxidizing* atmosphere today would destroy the molecules needed for chemical evolution. These scientists also point out that Jupiter and Saturn have *reducing* atmospheres, so it's likely a primitive Earth did, too.

On the other hand, there are those who insist that there is not much, if any, evidence for a primitive methane-ammonia atmosphere. P.H. Abelson, Director of the Carnegie Institute of Washington, said:

"The methane-ammonia hypothesis is in major trouble with respect to the ammonia component, for ammonia on the primitive earth would have quickly disappeared . . ."
([1]**Abelson,** 1966).

Rocks of the earliest ages should contain unusually large amounts of carbon or organic chemicals. This is not the case, says Abelson. Abelson believes in some kind of a reducing atmosphere, but geologist C. F. Davidson says there is no evidence that the Earth ever had an atmosphere much different from what it has now ([2]**Davidson**, 1963).

There is a growing consensus among geophysicists that the earth never had a methane-ammonia-hydrogen atmosphere. The current idea is that the early atmosphere consisted of carbon dioxide, nitrogen, and water vapor. There is, in fact, considerable evidence that the earth had a considerable amount of oxygen in its atmosphere, when the earliest sedimentary rocks were being deposited ([3]**Austin**, 1982). If so, this is fatal to all evolutionary origin-of-life theories.

Figure 2.1 New data show that the earliest atmosphere on earth may have been an oxygen atmosphere.

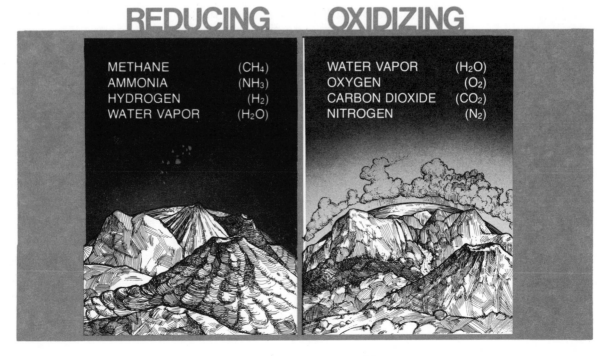

REDUCING   OXIDIZING

METHANE	(CH₄)
AMMONIA	(NH₃)
HYDROGEN	(H₂)
WATER VAPOR	(H₂O)

WATER VAPOR	(H₂O)
OXYGEN	(O₂)
CARBON DIOXIDE	(CO₂)
NITROGEN	(N₂)

The data below may help us decide how reasonable the *reducing atmosphere hypothesis* is. Which of the data below favors a reducing atmosphere?

1. Analysis shows that the atmosphere of Jupiter and Saturn contain methane and ammonia, and no oxygen.

2. Reduced chemicals are presently in the atmospheres of outer Jupiter-like planets, but the atmospheres of the inner earth-like planets (Mars, Venus) contain oxidized carbon and carbon dioxide (CO_2), instead of methane (CH_4).

3. Ammonia, present in the *reducing atmosphere*, is so soluable in water that on a planet like Earth, little would be in the air, and most would be dissolved in the oceans.

4. Ultraviolet light from the sun breaks down both ammonia and water vapor, releasing oxygen from the latter. Thus, any ammonia, if originally present, would disappear rapidly.

5. Some suggest that the early, atmosphere came from an outgassing of the interior of the earth. Volcanos are known to spew out CO_2 and H_2O in abundance.

6. Although some ancient iron ores are less oxidized at deeper levels, some of the oldest rocks known are in a highly oxidized state. Also, there are large deposits of sulfate minerals the highest state of oxidation of sulfur in supposedly very ancient rocks.

7. Without green plants to continually replenish the supply, very little, if any oxygen, in the air would exist free of chemical combination, but would probably combine with other atoms.

8. Biologically important molecules, such as the building blocks of DNA and proteins, would be destroyed by oxygen (O_2) in the early earth's atmosphere.

INTERPRET THE DATA

1. Does the circumstantial evidence we have in hand now establish the likelihood of a *reducing* atmosphere "beyond a reasonable doubt"? Or, does it conclusively disprove the hypothesis?

2. Do you think a *reducing* atmosphere for the early Earth is a *fact* that must be accepted, or more of an *assumption* made in support of the idea of chemical evolution?

Stage II

Simple Biological Molecules

Although scientists argue back and forth about the likelihood of a *reducing* atmosphere, they agree that energizing methane (CH_4), ammonia (NH_3) and water H_2O) does indeed produce a variety of biologically important molecules.

Stanley Miller's Experiment

Stanley Miller was one of the first to publish the results of such experiments. He placed all of the necessary ingredients in his apparatus (Figure 2.2) and let it operate continuously for about a week. Then the products that were in the trap were analyzed with the results described on the next page.

"A small number of relatively simple compounds accounted for a large proportion of the products. A very complicated mixture containing small amounts of a variety of compounds would have been anticipated. Furthermore, the major products . . . included a surprising number of substances that occur in living organisms" ([4]Miller and Orgel, 1974).

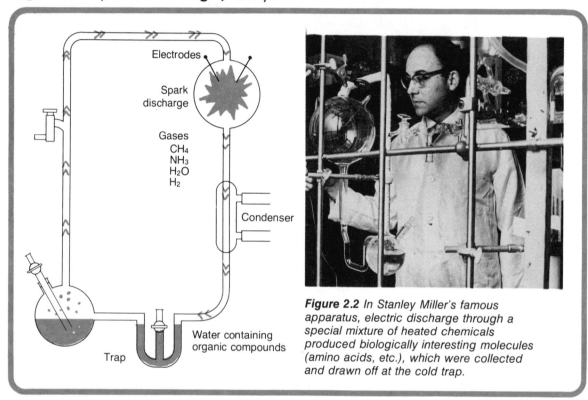

Electrodes

Spark discharge

Gases
CH$_4$
NH$_3$
H$_2$O
H$_2$

Condenser

Water containing organic compounds

Trap

Figure 2.2 *In Stanley Miller's famous apparatus, electric discharge through a special mixture of heated chemicals produced biologically interesting molecules (amino acids, etc.), which were collected and drawn off at the cold trap.*

REACTIONS

Reactions to Stanley Miller's Experiment
Stanley Miller's work received honored recognition in the 1950's. This was really the beginning of the modern study of chemical evolution.

First, the trivial nature of these results must be pointed out: All that were produced in these experiments were a few amino acids and other very simple compounds. The origin of life would require the production of **all** the amino acids, sugars, purines, pyrimidines, and other compounds, and their combination in the macromolecules, such as proteins, carbohydrates, DNA, and RNA, as well as the construction of the exceedingly complex, self-replicating cell.

A vital part of Miller's apparatus is a *cold trap* (Figure 2.2) to collect the products as they were formed from chemical reactions. Actually, without this trap, the chemical products would have been

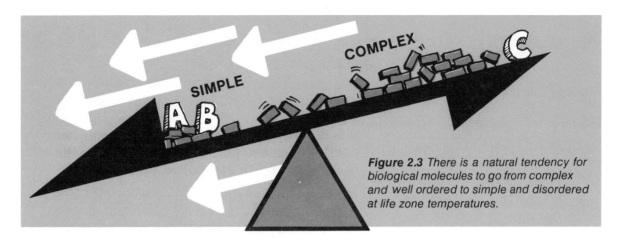

Figure 2.3 *There is a natural tendency for biological molecules to go from complex and well ordered to simple and disordered at life zone temperatures.*

destroyed by the energy source (electrical sparking). It is important for us to realize that the energy which forms the molocules is also the energy that *destroys* these molocules as they are formed. So, without the trap, a balanced (equilibrium between formation and destruction would be reached. Will this *balance* or *equilibrium point* favor formation (accumulating more complex molecules) or destruction of complex molecules (accumulating simpler molecules)? The biochemist tells us about this point:

equilibrium: a back-and-forth balanced end point where change in one direction equals change in the opposite direction so no net change occurs

> **"The equilibrium point in any chemical reaction is the point of lowest energy level. In a reaction involving the coupling of simple molecules to form more complex ones, the equilibrium point always lies toward that side which includes the simpler molecules . . ."**([5]**Gish**, 1972) (Figure 2.3).

The rate of destruction actually vastly exceeds the rate of formation.

Left to themselves and natural chemical processes, then, the biological molecules produced in Stanley Miller's apparatus could not accumulate for the next stage in chemical evolution without the chemist's cold trap. Could **tide pools**, lakes, or **metallic clays** serve as some sort of trap on the early Earth? Or could the early ocean have been near freezing? But solving the trap problem would make another problem, because the molecules that must be protected from energy sources also need that energy to advance to the next stage. Thus a trap actually would be fatal to any evolutionary theory.

Another point which must be considered is the

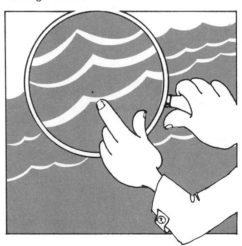

tide pools: pools that are formed from tides going in and out

metallic clays: clays rich in one or more metals

RIGHT MOLECULES IN THE WRONG PLACES

Figure 2.4 *Left to themselves, amino acids, sugars, and other biological molecules naturally react in destructive ways, and this makes some scientists wonder how such molecules in the early ocean could evolve toward life.*

concentration of the products in an ocean of water. D.E. Hull explains this dilemma from a physical chemist's viewpoint:

The physical chemist, guided by the proved principles of chemical thermodynamics and kinetics, cannot offer any encouragement to the biochemist [chemical evolutionist] who needs an ocean full of organic compounds to form even lifeless coacervates.

These estimates are not in conflict with Miller and others who have synthesized organic compounds with electrical discharges or ultraviolet light in the laboratory. They have merely used the well-known principle of increasing the yield of a reaction by selectively removing the product from the reacting mixture ([6]Hull, 1960).

Hull then went on to calculate the *equilibrium concentration* of some important biological molecules. The amino acid glycine, he said, would not likely reach over 10^{-12}M (one trillionth molar), and the sugar glucose 10^{-134}M (one molecule in a solution bigger than the universe!).

One final problem concerns the tendency of Miller's molecules to react in the "wrong" way. His experiment produced not only molecules *necessary* for life, but also a great many molecules *destructive* to life. He obtained not only short-chain, "left-handed" amino acids, for example, but also long-chain and "right-handed" amino acids that would interfere with further evolution. Furthermore, the sugars, if produced at all, would also be produced equally in the "left-handed" and "right-handed" forms, but only the "right-handed" forms will work in the DNA and RNA of living organisms. Ordinary chemistry cannot choose between "left-handed" and "right-handed" amino acids and sugars. Why then are only "left-handed" amino acids and "right-handed" sugars found in the proteins and nucleic acids, respectively, of living things today? In fact, no scientist has ever used the mixture of molecules in Miller's apparatus for the next step in chemical evolution, because, as one said, "The chemistry would be too hard."

A very interesting problem arises when sugars appear with amino acids. Placed together, they react with each other, and products that are not biological are formed. In fact, the following chemistry takes place: Since amino acids and sugars combine so readily, there would be few amino acids left to form *protein polymers,* and no sugars left to make the larger carbohydrates, or DNA and RNA. Biochemists frequently ask the question: "Since free amino acids and sugars, in effect, cancel each other out," then how could the necessary protein and DNA and other complex chemicals ever have arisen on a primitive earth?" Furthermore, calcium reacts with phosphoric acid to form the insoluble salt calcium phosphate. Since the supply of calcium in the earth's crust greatly exceeds that of phosphorus, all of the phosphorus on the hypothetical primitive earth would be removed and deposited on the bottom of the ocean as an insoluble rock.

The experiments of Miller and those following his lead definitely show how small biological compounds can be synthesized in the laboratory. But do they, in fact, show how such molecules could have been formed on a primitive earth? And do such molecules favor further evolution? These questions, and many others, face the experimenter concerned with life's origin.

INTERPRET THE DATA

1. **Have Stanley Miller and other chemical evolutionists succeeded in producing simple biological molecules by ordinary chemical means?**

2. **Why was the trap so important in Miller's experiment? Speculate about what might have served as a natural trap on a primitive earth. Is Miller's apparatus a reasonable model of what a primitive earth might have been like?**

3. **Besides producing molecules of biological importance, did Miller also produce molecules that might interfere with chemical evolution?**

4. **How do you think a creation scientist would interpret the results of Miller's experiment?**

5. **Make a chart showing both the *facts* and the *assumptions* involved in Miller's experiment, then try to decide what *conclusions* can be drawn. Do your *conclusions* favor the *evolution model*, the *creation model*, or leave the issue undecided? Why?**

Stage III

Complex Biological Molecules

Although the "Stage II" experiments of Miller, Ponnamperuma, and others are certainly interesting, the simple molecules produced are still little more than "dust of the ground" and a long way from life. The molecules produced are also formed in outer space without any connection to life, and the experiments simply do not explain the *organization* and *coordination* that makes living systems different from non-living.

Much of the coordination and organization of living systems depends on the complex biological molecules called **enzymes**. Enzymes are large *protein* molecules (chains of amino acids) coiled and folded into different shapes. The shape of the *enzyme* enables it to hold certain interlocking molecules in proper position for speedy reaction (Figure 2.5).

One example is the *enzyme*, catalase, that changes hydrogen peroxide (H_2O_2) into water and

enzymes: protein molecules that catalyze (speed up) specific chemical reactions

Figure 2.5 *Enzymes are protein molecules that cause rapid coupling and uncoupling of molecules with certain specific ("lock and key") shapes.*

oxygen. Looking at the structure of this *enzyme* (Figure 2.6), one finds that it includes *iron,* **heme**, and *protein*. A simple solution of *iron salt* (iron and chloride or bromide, etc.) shows some ability to decompose hydrogen peroxide. When *heme* is added to this mixture, however, the reaction is at least 1000 times more rapid. Now, when *iron, heme*, and the *specific protein* found in **catalase** are all used, the reaction speeds up several billion times.

In a living cell, *enzymes* act in *series* to get things done. In our muscle cells a "chain reaction" with about a dozen *enzymes* is required to obtain energy

heme: the deep red iron-containing group in hemoglobin ($C_{34}H_{32}N_4O_4Fe$).

catalase: the enzyme that catalyzes the conversion of hydrogen peroxide (H_2O_2) into water and oxygen

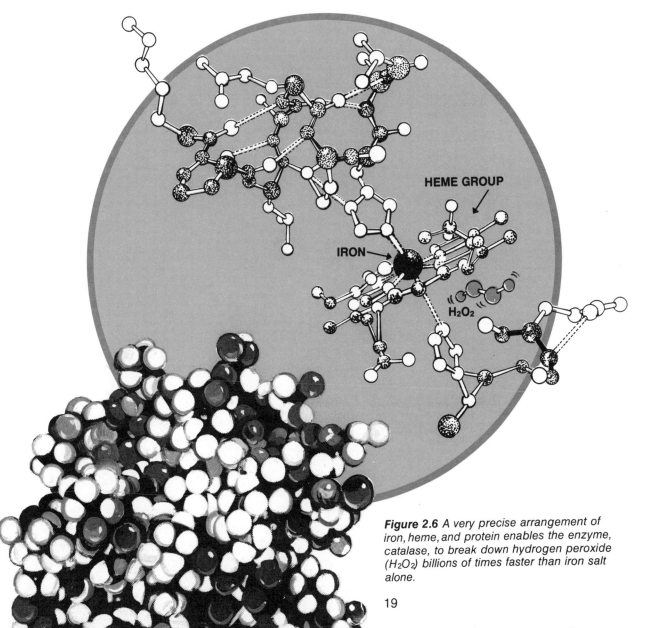

HEME GROUP

IRON

H_2O_2

Figure 2.6 *A very precise arrangement of iron, heme, and protein enables the enzyme, catalase, to break down hydrogen peroxide (H_2O_2) billions of times faster than iron salt alone.*

19

$C_6H_{12}O_6$

$C_6H_{11}O_6\text{-}(P)$

$C_6H_{11}O_6\text{-}(P)$

$C_6H_{10}O_6\text{-}(P_2)$

$C_3H_5O_3\text{-}(P)$

$C_3H_4O_4\text{-}(P_2)$

$C_3H_5O_4\text{-}(P)$

$C_3H_5O_4\text{-}(P)$

ATP

ATP

$C_3H_3O_3\text{-}(P)$

$C_3H_4O_3$

Figure 2.7 *It takes ten enzymes acting in a row to change sugar into pyruvic acid, and this only starts the process of harnessing energy for muscle contraction!*

random: lacking a definite plan or purpose

from changing sugar into lactic acid (Figure 2.7). Imagine this chain of events occuring over and over every time you move a muscle!

Enzymes are therefore an important key to the *coordination* of molecular reactions in living systems. For the investigator into life's origin, then, the question becomes, "Will time, chance, and natural chemical processes produce such *enzymes*?" Remembering always that *enzymes* are special kinds of *proteins*, let's look first at the formation of **random** *proteins* and then at the formation of *specific proteins*.

Sidney Fox and *Random Proteins*

Sidney Fox produced protein-like molecules by heating pure, dry amino acids at 150°-180°C for four to six hours. He thought something similar may once have happened on the edges of volcanoes ([7]Fox, 1988).

When he dissolved the product in hot water and allowed the solution to cool, he observed that small *microspheres* formed. These globules of **proteinoids** seemed to bud and grow, and Fox was encouraged to think he had discovered a possible pathway to the first living cell. The results sound exciting, but already questions may be buzzing in your mind. Write down the questions you think of before you turn this page.

proteinoids: protein-like molecules somewhat similar to natural proteins

Figure 2.8 Sidney Fox hypothesized that volcanic heat might cause pure amino acids to form proteins, and that rain would wash these proteins into pools before the heat could destroy them.

21

QUESTION THE DATA!

1. How could pure, dry amino acids collect on the early earth?

2. What do you think would happen if Fox left his amino acids at these temperatures for a longer period of time (over 10 hours)?

3. What problems would arise with amino acids on the edges of volcanoes that spew forth much (est. 70%) water? (Amino acids will break down if heated wet.)

4. Why didn't Fox start with a mixture of Miller's Stage - II molecules instead of pure, dry amino acids?

REACTIONS

Reactions to Fox's Experiments

A few scientists have honored Fox's work, but many are very skeptical about its relation to the true origin of life. Stanley Miller and Leslie Orgel made the following comments about this:

> If there were places where such polymerization could be accomplished, then it would still be necessary to show how the amino acids were brought to the lava and the peptides removed in an efficient manner ([4]Miller and Orgel, 1974).

> Dr. Fox's production of various protein-oids under such conditions, however, might . . . favor the destruction of any evolving life; catalysts [protein enzymes] only hasten the achievement of equilibrium without affecting the balance point of a reaction, and an organism at equilibrium is a dead organism" ([8]Parker, 1970).

Serine and threonine, important amino acids in proteins, are almost completely destroyed by the brutal heat. Furthermore, even if one begins with all left-handed amino acids, these brutal conditions convert amino acids to equal proportions of the right-handed and left-handed amino acids. These would be incorporated equally into proteins, and so the proteins so produced would contain random mixtures of these two forms, but the presence of a single right-handed form destroys all biological activity.

Specific Protein and Probability

There are 20 different kinds of amino acids, and the average protein consists of 300-500 such units in a chain. The number of possible *proteins* with only 200 amino acids per chain is 20^{200} or 10^{260} (the number 1 followed by 260 zeros)! The whole known universe contains far less than 10^{100} atoms, and 20 billion years is far less than 10^{20} seconds. So even if these *proteins* were as numerous as atoms in the universe, and even if a new set of *proteins* were produced every second for 20 billion years, the chances of finding our particular *protein* with 200 specified amino acids is still zero (1 chance in 10^{140})!

The Russian biochemist, A. I. Oparin, who is the "father" of the modern view of chemical evolution, was very conscious of this problem when he wrote,

the spontaneous formation of such an atomic arrangement in the protein molecule would seem as improbable as would the accidental origin of the text of Virgil's *Aneid* [a Latin epic poem] from scattered letter type ([9]Oparin, 1965).

INTERPRET THE DATA

1. Was Sidney Fox able to produce protein-like molecules using ordinary chemical means?

2. Do Fox's results argue for or against chemical evolution? Give reasons for your answer.

3. How would a creationist interpret Fox's experiment? Do you think any of the scientists quoted are creationists? If so, whom?

4. Mars has the same elements as Earth, many of the same molecules, and apparently once had liquid water. Have time, chance, and natural chemical processes produced any Stage III molecules on Mars? (You'll have to check the latest sources in your library for this one.)

Stage IV

Coacervates

If the development of Stage III polymers was possible on a primitive Earth, then Stage IV follows very easily. If you have watched fat droplets form, come together and divide in a bowl of soup, you can see there would be a natural tendency for Stage III molecules to form similar gel-like globs, called *coacervates*. Fox's *microspheres* are also a kind of coacervate. *Oparin* believes that coacervates may have become more and more complex by absorbing chemical compounds when left under the right conditions for a period of time. He believes there is a possibility that a living cell could have formed this way (Figure 2.9).

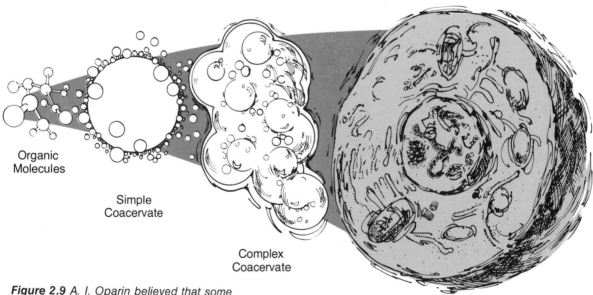

Organic
Molecules

Simple
Coacervate

Complex
Coacervate

First Simple
Living Organisms

Figure 2.9 *A. I. Oparin believed that some coacervate globule might eventually absorb just the right molecules to become a living cell, although he knew that coacervates are unstable and would also absorb harmful molecules.*

24

In his experiments with coacervates, Oparin found that he could place active *enzymes* in a coacervate solution and they would continue to function chemically. Coacervates are certainly interesting. Here are some other observations about them:

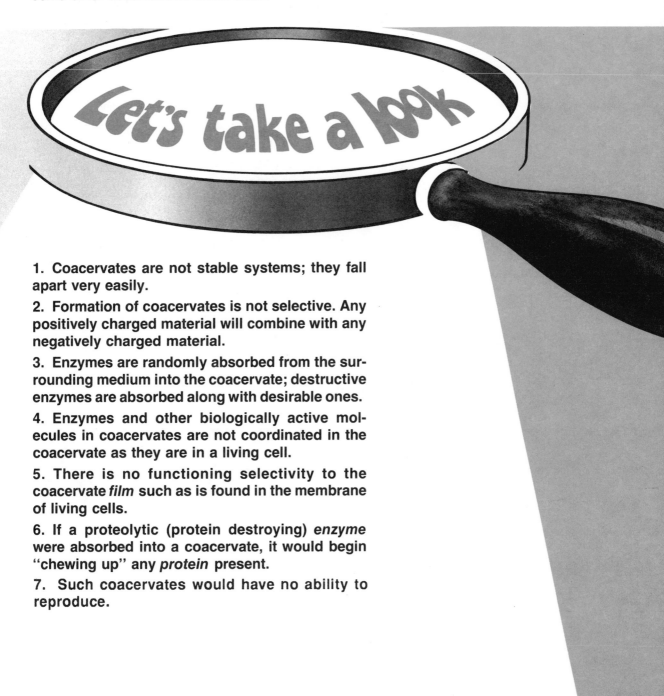

1. Coacervates are not stable systems; they fall apart very easily.

2. Formation of coacervates is not selective. Any positively charged material will combine with any negatively charged material.

3. Enzymes are randomly absorbed from the surrounding medium into the coacervate; destructive enzymes are absorbed along with desirable ones.

4. Enzymes and other biologically active molecules in coacervates are not coordinated in the coacervate as they are in a living cell.

5. There is no functioning selectivity to the coacervate *film* such as is found in the membrane of living cells.

6. If a proteolytic (protein destroying) *enzyme* were absorbed into a coacervate, it would begin "chewing up" any *protein* present.

7. Such coacervates would have no ability to reproduce.

INTERPRET THE DATA

1. **What is the basic difference between a coacervate and a living cell? Speculate on how a coacervate could become a living cell by evolution.**

2. **Do you see any problems in going from *polymers* to coacervates to living cells? Explain.**

3. **Coacervates certainly form readily by natural processes. How do you think creation scientists would respond to this evidence?**

Stage V

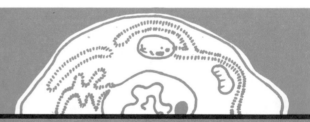

The First Living Cells and DNA

The biggest difference between coacervates and the first living cells is *coordination* of the molecules involved. A.I. Oparin wrote:

respiratory: obtaining and using oxygen in the cellular process that generates energy from food

> **If, for instance, one was to prepare an artificial mixture of all the enzymes which promote the separate reactions constituting the *respiratory* process, he would still fail to reproduce respiration by means of this mixture...for the simple reason that the reaction velocities will not be properly and mutually coordinated"** ([9]Oparin, 1965).

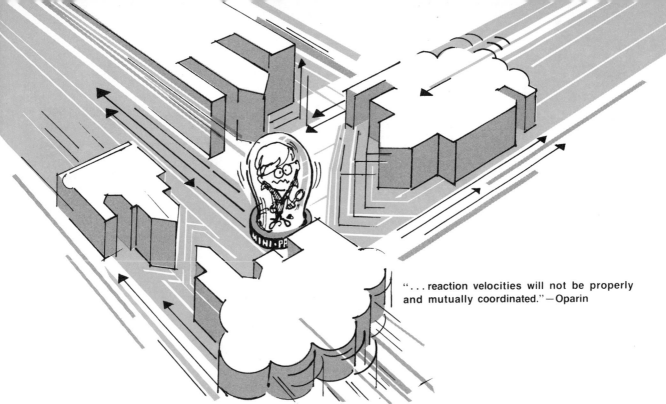

"...reaction velocities will not be properly and mutually coordinated." —Oparin

In all living systems as we know them today, the most fundamental example of coordination is the way DNA codes for proteins. DNA is the molecule of heredity, and proteins are the molecules of structure and function.

Guanine (G), cytosine (C), adenine (A), and thymine (T) are the four chemical groups (bases) in DNA that act like alphabet letters "spelling out" hereditary instructions. The series of *bases* in a DNA molecule (or gene) are used to line up the series of *amino acids* that make each protein capable of performing a certain biological task. In this DNA code, a triplet group (three bases) specifies each amino acid. For example, the amino acid sequence of proline-valine-glutamic acid can be specified by the DNA code G-G-A (proline code), C-A-T (valine code), and C-T-T (glutamic acid code).

The synthesis of proteins not only requires the DNA code but also requires many specific enzymes, which are proteins themselves, and several **RNA** molecules to select and activate each amino acid. In fact, a very

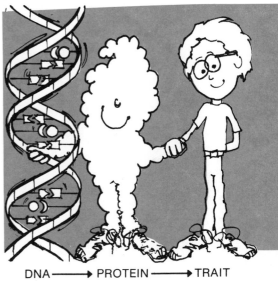

DNA ⟶ PROTEIN ⟶ TRAIT

RNA: a nucleic acid similar to DNA, but with the sugar ribose (vs. deoxyribose) and the base U (vs. T).

YOU HAVE INHERITED A SMALL NOSE
CCAGTTACCGTAAGCTTCCGATAATCGC

Figure 2.10 The bases in DNA "spell out" instructions for producing protein molecules and their related hereditary traits.

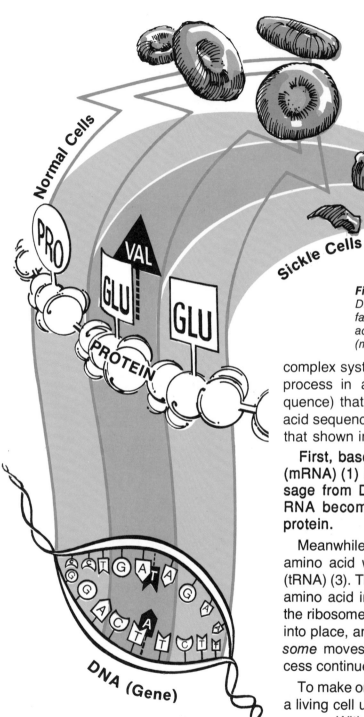

TRAITS

Normal Cells

Sickle Cells

Figure 2.11 *A seemingly minor change in DNA (T in place of A) causes cells to produce faulty hemoglobin (valine in place of glutamic acid), and the result is sickle cell anemia (misshapen red blood cells).*

complex system is required. Let us look at the whole process in a diagram. Imagine a DNA (base sequence) that codes for a particular protein (amino acid sequence) beginning the process something like that shown in Figure 2.12.

First, based on the DNA code, messenger RNA (mRNA) (1) is produced. This mRNA carries a message from DNA to the **ribosomes** (2), where the RNA becomes a template for making a certain protein.

Meanwhile, specific *activating enzymes* join each amino acid with its specific transfer RNA molecule (tRNA) (3). The properly coded tRNA locks its specific amino acid into place along the mRNA template on the ribosome (4). When another amino acid is locked into place, an enzyme connects them. Then the *ribosome* moves to the next mRNA codon, and the process continues until a complete protein is formed (5).

To make one protein according to DNA instructions, a living cell uses over 70 specific proteins and much energy. Without a constant flow of amino acids and energy molecules (ATP) directed by very specific enzymes, the whole process would break down, just as it does at death. Could time, chance, and natural chemical processes coordinate all these molecules in the ways needed to produce the first living cell?

ribosomes: tiny, dense granules in the cell where proteins are produced through coordinating the action of messenger and transfer RNA, enzymes, and energy (ATP)

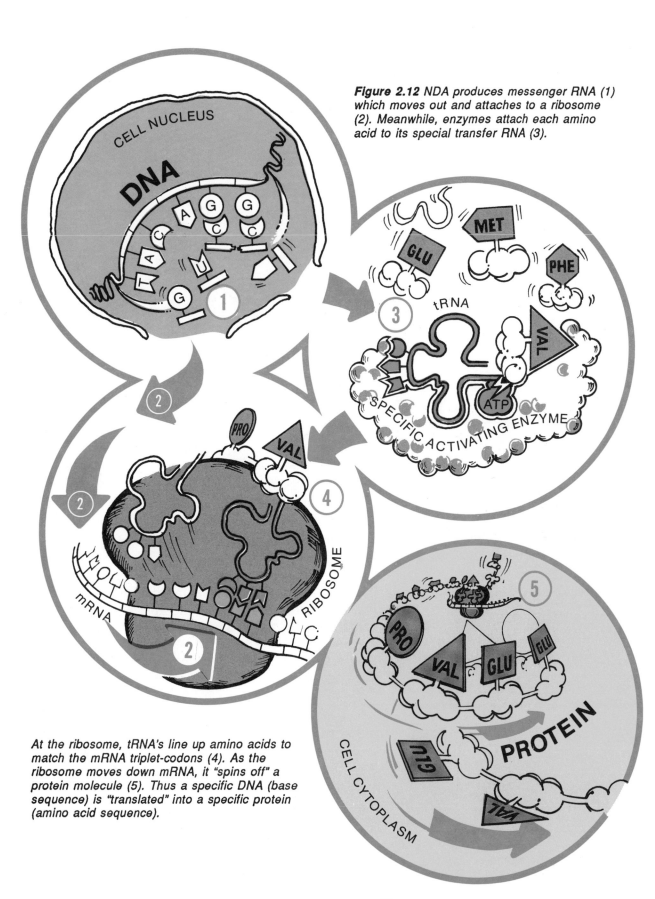

Figure 2.12 NDA produces messenger RNA (1) which moves out and attaches to a ribosome (2). Meanwhile, enzymes attach each amino acid to its special transfer RNA (3).

At the ribosome, tRNA's line up amino acids to match the mRNA triplet-codons (4). As the ribosome moves down mRNA, it "spins off" a protein molecule (5). Thus a specific DNA (base sequence) is "translated" into a specific protein (amino acid sequence).

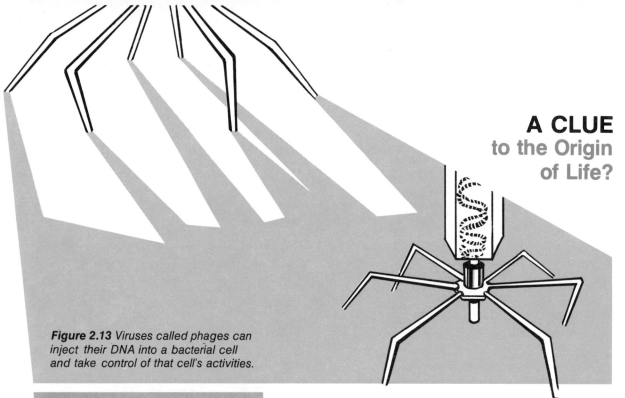

A CLUE
to the Origin of Life?

Figure 2.13 Viruses called phages can inject their DNA into a bacterial cell and take control of that cell's activities.

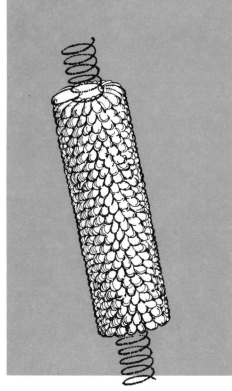

Figure 2.14 The tobacco mosaic virus consists of a long RNA molecule wound up inside a cylinder of 2130 identical protein molecules.

metabolism: a sum of all the chemical processes of the body

Some people hoped viruses would provide a clue to the origin of the first living cell. A virus, basically, is a protein covering a nucleic acid (either RNA or DNA). The virus attaches to a cell and its DNA or RNA is injected. When the virus DNA gets into the cell, it takes over the genetics of the cell. The virus, in effect, causes the cell to reproduce more virus DNA; to manufacture virus protein for its outer coat; and to assemble a complete virus.

Can the virus be a "missing link" between the living and nonliving worlds? Some scientists once thought so, but now they know that viruses can only be produced by living cells. So it seems that living cells with their own energy-harnessing **metabolism** must have come before viruses. Besides, viruses already have the link between DNA and protein established, so they really don't help us to understand the origin of this special relationship.

Many notable scientists have wondered about the possibility that time, chance, and natural chemical processes could ever produce the relationship we see in living cells between DNA and protein. Consider their comments and relate them to the virus:

1. Francis Crick, who shared the Nobel prize for the discovery of DNA's structure, said that the "trick" in life is to make the translation machinery; and he added that you have to have an "elaborate biosynthetic mechanism" to make a 4-letter code (DNA) into a 20-letter code (protein).

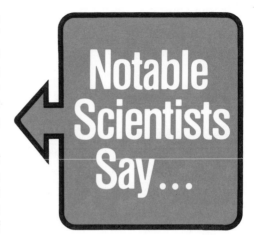

Notable Scientists Say...

2. Leslie Orgel, co-author with Stanley Miller of the "Origins of Life on the Earth," said the "big problem" is to tie DNA reproduction into the machinery to make protein; and he openly wondered, "Is the problem chemical or philosophical?" That is, can future discoveries in chemistry bridge this "big gap" in the story of evolution, or must we look beyond chemistry to the realm of ideas in philosophy?

3. Dr. O'Connor, in his college chemistry textbook writes: "There is no reason for scientists to discard belief in God. Indeed, there are many who feel the need for a faith that suggests human life is more than a series of chemical changes....If true, and many of us are convinced that it is, it is far more important than anything science, technology, or this world has to offer" ([10]**O'Connor**, 1974).

INTERPRET THE DATA

1. Is there good evidence that some of the large and complex molecules needed for life on Earth can be produced by time, chance, and natural chemical processes, under assumed primitive, earth conditions?

2. Do we have any good evidence that these molecules can separate from harmful ones and spontaneously organize themselves into coordinated living systems?

3. Can viruses be regarded as a link between life and non-life. Why or why not?

4. Carl Sagan, a professor at Cornell University, once wrote that "a fairly straightforward organic chemistry yields complex molecules that combine into self-replicating [reproducing] molecular systems: the first terrestrial organisms [the first living cells on Earth]." Is his statement based on data and experimental evidence, or faith in future discoveries he hopes chemical evolutionists will make?

CONCLUSION:

The Evolutionary Scenario

You have now come a long way in your investigation of life's origin. Certainly the evolutionary scenario captures the imagination and stimulates much thinking. Let's review the whole story now, as you ask yourself: What are the facts? . . . the assumptions? . . . the arguments pro and con?

According to the evolution model, the story of life's origin began with high energy bombardment of methane (CH_4) and ammonia (NH_3) in the early Earth's atmosphere (Stage I). Amino acids, sugars, and other simple molecules thus formed rained down into the oceans (Stage II). Perhaps on the edges of volcanoes, these molecules combined to form large *polymers*, somewhat like the DNA and protein molecules we have today (Stage III). In *tidepools* and along shorelines, these *polymers* formed globs or coacervates that began to absorb smaller molecules (Stage IV). Then purely by chance made likely by billions of years of time, at least one of these coacervates absorbed the right molecules for harnessing energy and for making proteins from DNA code, and life began (Stage V). The chain of life now leads all the way up to man, the first organism to contemplate his own evolution and to seize control of his own destiny.

A stirring story, and one with important implications for understanding life, but . . .

Pure amino acids must be heated to 175°C on dry area of

EVALUATE THE MODEL

1. What do you think are the major *assumptions* involved in the evolution model?

2. What do you think are the strongest data *supporting* the evolution model?

3. What are the strongest arguments *against* the model?

4. Mars has many of the same elements and compounds as Earth and has presumably been around as long, yet has no traces of life or even of biological molecules. Is this important to know while evaluating the evolution model?

5 Is it fair to say that the case for chemical evolution is established "beyond a reasonable doubt"? Or, do you think the data casts enough doubt on the model that alternatives should be explored?

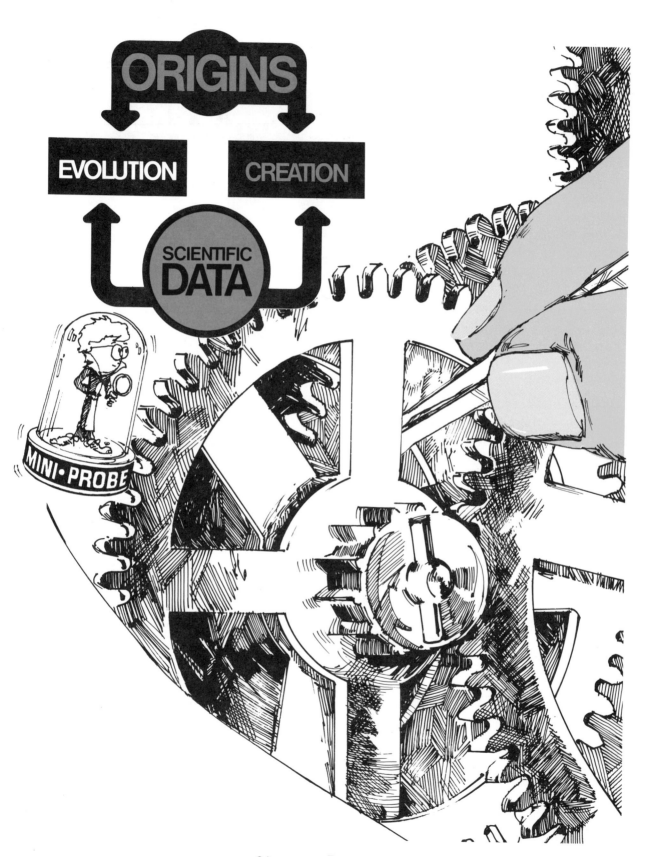

CHAPTER THREE

Creation Model for the Origin of Life

According to the *creation* model, watching the origin of life on Earth would be something like watching a master craftsman put together a finely-tooled grandfather clock. His creative design and careful coordination give the bits of wood and metal and glass properties of time-keeping and beauty that they could never develop on their own. Indeed, the clock will need his continuing care. Time, chance, and natural processes will tend to wear out the clock, so that it will need repeated cleaning, oiling, and winding, for example. In a similar way, living systems are molecular pieces whose property of life is also a product of creative design and craftsmanship. Living cells, too, need a constant energy supply and repair to keep themselves from the ravages of time, chance, and harmful chemical reactions.

The cell is the smallest unit of a living system. Multiplied millions of dollars have been spent to find the key to life that seems to be hidden within the membrane of the living cell. Each time we get closer to the secret of life, it seems that its orderliness and complexity point us more clearly toward a created origin.

The living *bacterial cell* is only about 1/1000 of a millimeter across, yet the bacterium has often been called a chemical factory. In fact, its complexity continues to boggle the minds of scientists. Green and Goldberger have aptly stated that: "The macromolecule to cell transition lies beyond the range of testable hypotheses. We have no basis for postulating that cells arose on this planet" ([11]**Green** and **Goldberger,** 1967).

Figure 3.1 *A bacterial cell is like a complex chemical factory regulated by DNA, organized by membranes, and run by enzyme and energy molecules.*

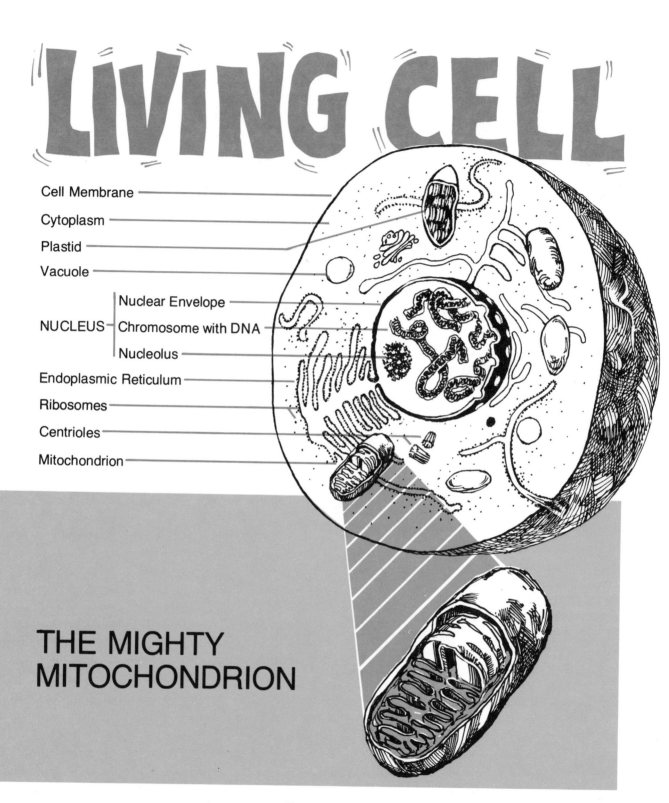

LIVING CELL

Cell Membrane

Cytoplasm

Plastid

Vacuole

NUCLEUS
- Nuclear Envelope
- Chromosome with DNA
- Nucleolus

Endoplasmic Reticulum

Ribosomes

Centrioles

Mitochondrion

THE MIGHTY MITOCHONDRION

Figure 3.2 *Most living cells include many complex parts, all working together. Cellular respiration within mitochondria, for example, supplies most of the energy for other cell processes.*

The simplest living cell with a nucleus contains many **organelles**, like the **mitochondrion**, that are each as complicated as a bacterial cell (Figure 3.1). The marvelously intricate nature of living cells is enough to make many scientists look beyond the molecules themselves for an explanation of life's origin. Still, creationists say that it is the *kind of* design we observe, not just the complexity, that provides evidence of *creation.* According to the creation model, knowledge we presently have of living systems suggests that:

organelles: small, specialized, organ-like structures present in the cell

mitochondrion: the "powerhouse of the cell," i.e., the specialized structure (organelle) where energy is harnessed from food using oxygen in the respiratory process

1. Time and chance would make the evolutionary origin of life impossible.

2. Natural chemical processes would work against the origin of life unless selected, integrated, and controlled by creative intelligence.

3. The kind of relationships we see among molecules in living systems is like the kind of relationships we see in objects created by man, where organization gives the system properties the molecules themselves do not have.

Remember the rock formation example in Chapter One (page 3)? Time, chance, and natural processes of weathering and erosion would produce the kind of relationship seen in the spontaneous formation; but time, chance, and natural processes would act to erase the kind of relationship seen in the sculptured or created formation. You may want to review that example as you sharpen your detective's wits once again, and take a look at the facts.

Both evolutionists and creationists have the same data or the same circumstantial evidence, but who has the better interpretation? Can the case for *creation* be established "beyond a reasonable doubt"? Is the creationist's case as good, better, or worse than the evolutionist's? Nobody can tell you the answer to these questions. Look carefully at the following evidence and see if you can decide for yourself.

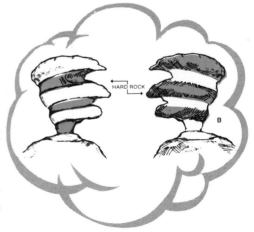

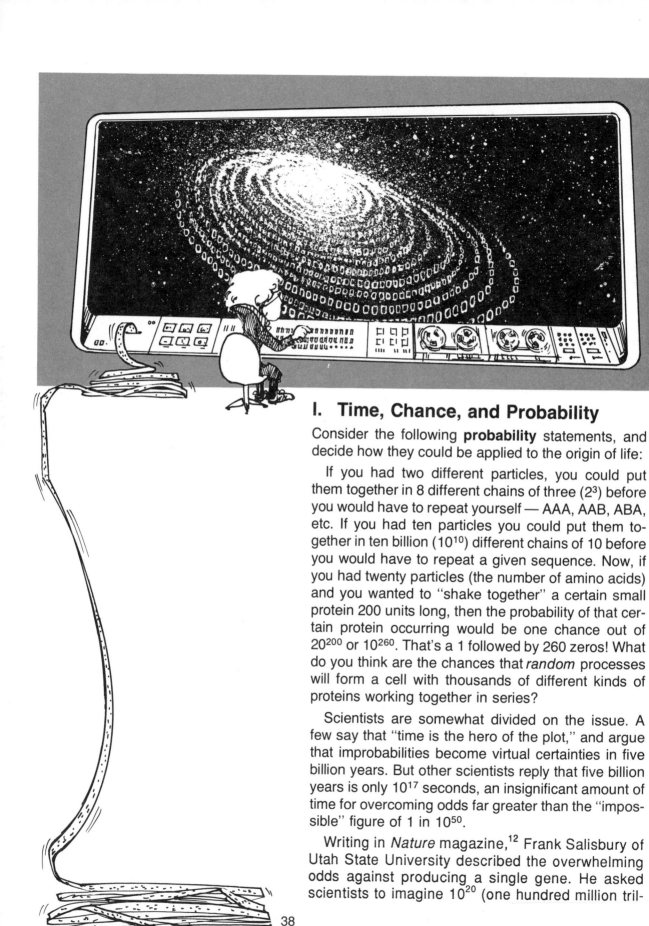

I. Time, Chance, and Probability

Consider the following **probability** statements, and decide how they could be applied to the origin of life:

If you had two different particles, you could put them together in 8 different chains of three (2^3) before you would have to repeat yourself — AAA, AAB, ABA, etc. If you had ten particles you could put them together in ten billion (10^{10}) different chains of 10 before you would have to repeat a given sequence. Now, if you had twenty particles (the number of amino acids) and you wanted to "shake together" a certain small protein 200 units long, then the probability of that certain protein occurring would be one chance out of 20^{200} or 10^{260}. That's a 1 followed by 260 zeros! What do you think are the chances that *random* processes will form a cell with thousands of different kinds of proteins working together in series?

Scientists are somewhat divided on the issue. A few say that "time is the hero of the plot," and argue that improbabilities become virtual certainties in five billion years. But other scientists reply that five billion years is only 10^{17} seconds, an insignificant amount of time for overcoming odds far greater than the "impossible" figure of 1 in 10^{50}.

Writing in *Nature* magazine,[12] Frank Salisbury of Utah State University described the overwhelming odds against producing a single gene. He asked scientists to imagine 10^{20} (one hundred million tril-

38

lion) planets each with an ocean 2 km deep. Each ocean is fairly rich in gene-sized DNA fragments, he supposes. Each DNA reproduces a million (10^6) times per second, and a mutation occurs at each reproduction. Under such favorable conditions, Salisbury calculates it would take trillions of universes to have much chance of producing a single kind of gene in four billion years—even if 10^{100} different DNA molecules could serve the same gene function! "Impossible" is almost too mild a term for such an event, and living systems require much more than one gene!

INTERPRET THE DATA

1. **Try these thought experiments:**

    a. Put five colored objects in a sack, and offer a friend $5 if he can pull out the five colors in a certain order, returning the object after each choice. (The odds here are only 1 in 3125, so if you're worried about your $5, make it $7 and 7 colors in a row of 7, and the odds are nearly 1 in a million in your favor!)

    b. Now, offer your friend any amount of money you want to get two sequences in a row; the odds of two specific five-color sequences in a row are only about 1 in ten million! (Ask your math teacher about the law of compound probabilities.)

    c. How do you think these simple experiments relate to the origin of life?

2. **Try this thought experiment: Put pieces of a watch in a bag (magnetized, if you wish) and begin shaking. How long would you have to shake before you would have a reasonable chance of getting all the pieces together and ticking? Is this a fair analogy to the problem of getting proteins, DNA, and other molecules together in a "working combination" (life)? Why or why not?**

2nd LAW
of Thermodynamics

II. Chemical Processes and Thermodynamics

What do molecules do when left "on their own" over long periods of time? The answer involves the *Second Law of Thermodynamics* and *kinetics*. According to the Second Law, molecules naturally react in ways that *decrease* the useful energy available for work ("free energy") and increase the overall disorder ("entropy") in a system.

The effects of the Second Law are seen in many everyday processes: heat flowing from warm to cold objects, water flowing downhill, automobile engines wearing out, the breakdown of meat proteins in boiling water, soft rock wearing away faster than hard, and dead bodies decomposing. All these are examples of energy and order running downhill to the lowest level (*equilibrium*) for the temperature involved. Does this remind you of what happens to your room as a result of time and chance and "natural processes"?

But don't feel too bad! The Second Law does permit an *increase* in order and complexity — *IF* you are willing to pay the price and can satisfy certain conditions. After all, acorns grow into oak trees, and a car can take you from one place to another! But what makes your car work? You need a source of energy (the gasoline) continuously supplied, an engine which can harness that energy for a particular purpose, and someone to drive the car! The acorn, too, has its

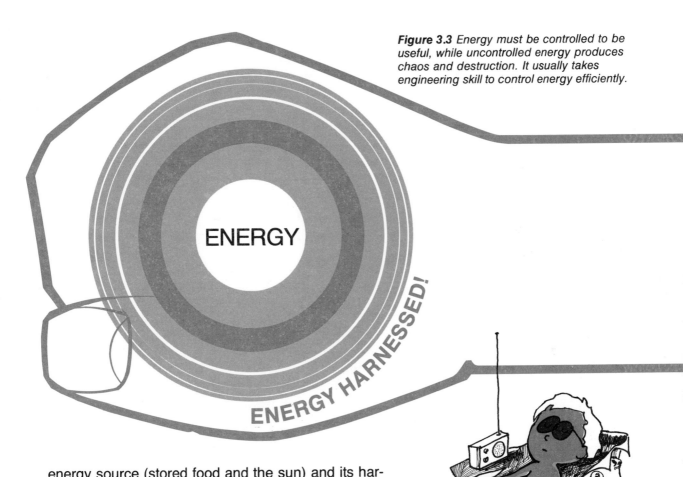

ENERGY

ENERGY HARNESSED!

Figure 3.3 Energy must be controlled to be useful, while uncontrolled energy produces chaos and destruction. It usually takes engineering skill to control energy efficiently.

energy source (stored food and the sun) and its harnessing system (DNA, *photosynthetic* enzymes, etc.) Even your room that gets messy so easily can be cleaned up, but it takes a lot of work to put things in their proper places.

But what does energy do without a harnessing system? What do gasoline explosions do when not controlled by the cylinder and piston of a car? An oak tree can use the sun's energy, but what would lying in the sun all day do to you? More to the point, what would energy sources such as heat, ultraviolet, and **high-energy phosphates** like ATP do to a *random* collection of molecules on a primitive Earth?

Creationists argue that the principles of **thermodynamics** and **kinetics** make chemical evolution impossible. The basic belief of evolutionists is that time, chance, and natural chemical processes will produce life without any "outside help." But creationists say evolutionists are inconsistent, because evolutionists must reject the laws of *thermodynamics* and *kinetics* to make evolution work.

high-energy phosphates: molecules like ATP (adenosine triphosphate) that boost the energy level of other molecules by giving them phosphate groups (H_2PO_4-)

thermodynamics: the science dealing with heat flow and the efficiency of energy transfer and exchange

kinetics: the science dealing with rates of change in a chemical system

How do scientists favoring evolution respond?

1. Many evolutionists simply say that there is no problem; that energy from the sun (disintegrating at 4½ million tons per second!) provides enough energy for the evolution of life on Earth; and they assume a harnessing system will eventually appear.

2. Ilya Prigogine, who recently received a Nobel prize for his work in non-*equilibrium thermodynamics*, believes the problem is very real, but that random fluctuations in energy levels and autocatalysis might solve the problem. But he admits that he must assume a continual flow of energy and specific raw materials to maintain his "dissipative structures."

3. Albert Szent-Gyorgi, biophysicist and also winner of a Nobel prize in science, feels there must be some as yet undiscovered "law of syntropy" that makes molecular organization move "uphill" against the effects of the Second Law (the "law of entropy").

INTERPRET THE DATA

1. According to the Second Law, is the spontaneous appearance of ordered structures im-

possible? If not, what must you have to get order from disorder?

2. Ice crystals are beautiful, ordered structures that form spontaneously when water cools down. As the crystal forms, it uses up the energy supplied by the partial electric charges on the molecules, so the crystal is incapable of further change until more energy is added. How might an evolutionist interpret crystal formation? . . . . a creationist?

3. According to the Second Law, it's easier to tear down an ordered system than it is to build it up. Can you give some everyday examples of this Law in action? How does this apply to the origin of life?

4. Although Szent-Gyorgi is an evolutionist, is his search for the "law of *syntropy*" consistent with the basic assumptions of chemical evolution? Why or why not? How would a creationist interpret his remarks?

III. Kind of Design

Suppose you were hiking through some back trails in the mountains and found a simple fragment of glazed clay with a pattern of stains like the one in Figure 3.4. Without hardly thinking about it, you would conclude that the object was the product of intelligent design. Although you had not seen the pottery produced, you might suppose the creator to be an Indian.

A pottery fragment certainly isn't complex or mysterious. Yet you can identify it as a created object. Why? Because the *kind* of design must have been imposed from the outside. It's not the kind of design

Figure 3.4 *Archaeologists use bits of pottery to infer the existence of past civilizations, so biologists should be free to infer from our observations that living systems are also the product of intelligent creation.*

produced by time, chance, and natural processes. So it's the *kind* of design, not design by itself or complexity, that really provides evidence of creation. Even without seeing the creator or the act of creation, you can recognize a created object by the kind of relationships observed.

The complexity of the DNA-protein relationship is sometimes used as an argument against chemical evolution (pp. 26-31). Their *kind* of relationship, in its basic simplicity, can be used as evidence for creation.

DNA is a chain of nucleotides with bases sticking out along it. A protein is a chain of amino acids with variable **"R" groups** sticking out along it. In all living systems, a chain of bases (three at a time) is used to make a chain of *"R" groups.* Is there any chemical basis for this reaction? None. All the *natural reactions*, such as base-acid, act to *break down* the base-*R-group* relationship.

To have a chain of bases make a chain of *"R" groups*, order has to be imposed from the outside. The molecules that impose this order (various enzymes and RNA molecules) are *not* simply speeding up chemical reactions that would happen anyway without them. They are establishing a new pattern that has no basis in the molecules themselves.

Look again at the way living cells use DNA to direct production of proteins (see Figure 2.12, page 29). *Ribosomes* break the messenger RNA code up into triplets, a logical "language grouping" that has nothing to do with the natural chemistry of the bases at all. Then look at the activating enzymes in the same figure. Near center is a natural pairing of transfer RNA with activated amino acids, but this reaction by itself would foul up the coding. So these activating enzymes also impose a relationship between triplet bases and *R groups* that is more like the relationships imposed in stone by human craftsmen and sculptors than of patterns in stone cut by natural processes of weathering and erosion.

In that sense, says the creationist, life is something like a television set. No laws of physics are violated in the *operation* of the TV set, but those laws don't explain the *origin* of the TV set. The "missing ingredient" is *not* some mysterious "vital force." The missing ingredient needed to explain

R groups: the distinctive portions of amino acids, which are lined up along a protein polymer like bases are lined up along DNA

Figure 3.5 This illustration shows a segment of a DNA molecule.

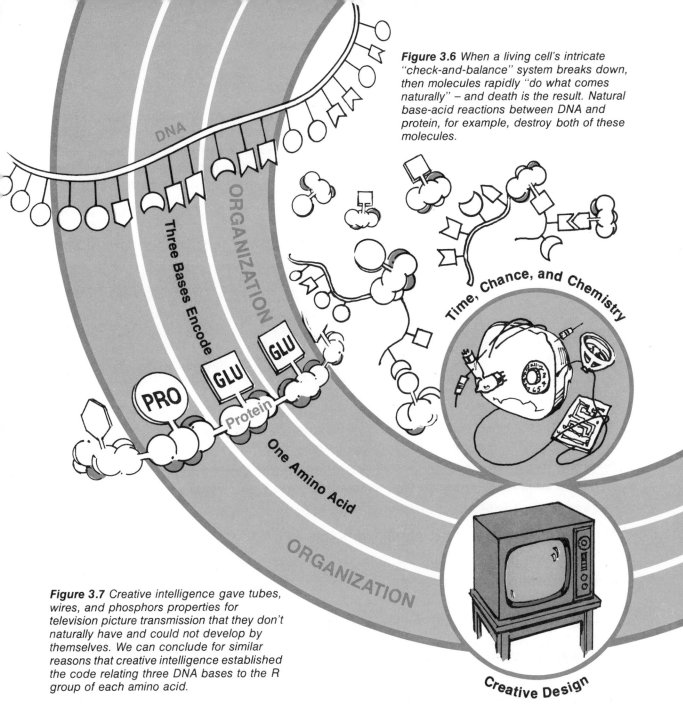

Figure 3.6 When a living cell's intricate "check-and-balance" system breaks down, then molecules rapidly "do what comes naturally" – and death is the result. Natural base-acid reactions between DNA and protein, for example, destroy both of these molecules.

DNA

ORGANIZATION

Three Bases Encode

Time, Chance, and Chemistry

PRO GLU GLU

Protein

One Amino Acid

ORGANIZATION

Creative Design

Figure 3.7 Creative intelligence gave tubes, wires, and phosphors properties for television picture transmission that they don't naturally have and could not develop by themselves. We can conclude for similar reasons that creative intelligence established the code relating three DNA bases to the R group of each amino acid.

the origin of the TV set is simply *organization* – the scientifically understandable and useful application of intelligent creative design. So it is with life. The molecules operate in living systems so that they can "multiply after their kinds," but organization in living systems had to come from an external creative agent.

Consider the matter of information, or codes. Can meaningful information or codes be generated spontaneously within simple systems, or must they be introduced by intelligent sources from the outside? Consider, for example, a source of dots and dashes. These might, by chance, be arranged in the sequence • • • — — — • • • . This would, of course, by itself, be meaningless, and thus would convey no information. If, however, intelligent human beings have invented the Morse code, and one knows Morse code, than • • • — — — • • • conveys the information SOS. Again, however, SOS, by itself, conveys no specific information unless intelligent human beings have previously agreed that SOS means "**S**ave **O**ur **S**hip" (Send help!). So even if a DNA molecule could be synthesized on the hypothetical primitive earth, this DNA molecule would contain no useful information unless there was a pre-existing system to utilize that information in a meaningful way.

Everyone can see the evidence of purposeful design in the works of man. In the same way, creationists say that observations and logical inferences from scientific data provide evidence of a creator's design in living systems.

EVALUATE THE MODEL

1. Some scientists say that you can't have evidence of creation without being able to see the creator. Do you agree? Why, or why not?

2. "It is not design itself nor complexity of design, but *kind* of design that provides evidence of creation." Explain this statement, using an example.

3. What are the basic *assumptions* of the creation model?

4. What do you think are the strongest points *in favor* of the creation model?

5. What do you think are the strongest points *against* the creation model?

6. Can the evolution and creation models both be inferred from scientific data? Can scientific analysis settle the issue?

7. Which view, evolution or creation, seems to have the greatest need for further research to establish its claim, and which seems to rest on concepts and observations we already have in hand?

8. If man created "life in a test tube," would this provide evidence of evolution, or creation, or would it leave the matter unsettled?

9. Which model depends most on assumptions and religious or philosophical preconceptions? Contrast the basic assumptions of each model.

10. You have now looked at both the case for evolution and the case for creation in this important matter of life's origin. Can either model be established or refuted "beyond a reasonable doubt?" Which model do *YOU* think fits the data best — and why?

CONCLUSION

One of the most challenging questions to face man-kind has been addressed in this book. Perhaps the clearest point is that answers cannot be found through scientific study alone. Nevertheless, you now have some idea of the arguments for the evolution model as well as arguments for the creation model. Decisions, if made at all, must be made from the standpoint of " Which model fits all the data best, taken as a whole?"

You have heard the best of numerous arguments and proposals put forth by evolutionary thought as well as creation thought. The opportunity to make decisions about these views lies in your hands. No one should tell you that one or the other is the better argument without putting forth scientific evidence to support his views. This study has openly discussed both sides of the question; now it becomes the data that can influence you one way or the other. The question is yours — which model fits these data best?

But are the scientific data the only relevant data? Science is only one approach to knowledge. It is best at helping us to understand observable, repeatable events in our world today. Data on origins lie outside this domain. As you have seen, the scientific data is only circumstantial and subject to more than one interpretation. Should the data of philosophy, religion, and revelation also be consulted? Is it fair to talk about the existence of "God"? A thoughtful person can't really avoid these questions, although they make it hard to study origins objectively. We hope this book will help you to understand better the reasons for your own choice, and to treat with respect and compassion the choices of others.

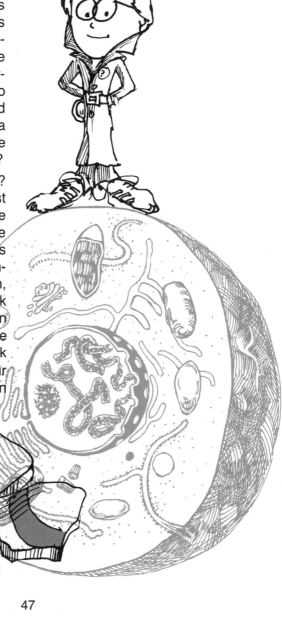

47

RESOURCE BOOKS

Evolution Books

1. **Gastonguay, Paul R.,** *Evolution for Everyone,* Bobbs Merrill Co., Inc., Indianapolis, 1974 (Biological Science Curriculum Study), 213 pp.

*2. **Marquand, Josephine,** *Life: Its Nature, Origins and Distribution,* W.W. Norton & Company, Inc., New York, 1971, 96 pp.

*3. **Hodge, Paul W.,** *Concepts of the Universe,* McGraw-Hill Book Company, New York, 1969, 125 pp.

*4. **Darwin, Charles,** *Introduction to Origin of Species,* Everyman's Library, E. P. Dutton and Co., Inc., New York.

*5. **Blum, Harold F.,** *Time's Arrow and Evolution,* Princeton University Press, Princeton, NJ, 1968, 232 pp.

6. **Oram, Raymond F.,** *Biology: Living Systems,* Charles E. Merrill Pub. Co., Columbus, OH, 1973, 784 pp.

*7. **Day, M.H.,** *The Fossil History of Man,* Oxford University Press, Ely House, London W.I., 16 pp.

8. **Miller, S.** and **Orgel, L.,** *The Origins of Life On the Earth,* Prentice-Hall, Inc. NJ, 1974

Creation Books

1. **Morris, Henry M.,** *Scientific Creationism,* Creation-Life Publishers, San Diego, 1974, 217 pp.

2. **Gish, Duane T.,** *Evolution: The Challenge of the Fossil Record,* Master Books, San Diego, 1985.

*3. **Barnes, Thomas G.,** *Origin and Destiny of the Earth's Magnetic Field,* Institute for Creation Research, San Diego, 1973, 64 pp.

*4. **Gish, Duane T.,** *Speculations and Experiments on the Origin of Life,* Institute for Creation Research, San Diego, 1972, 41 pp.

*5. **Lammerts, W.E.,** Ed., *Scientific Studies in Special Creation,* Presbyterian and Reformed Publishing Co., Philadelphia, PA, 1971, 343 pp.

*6. **Slusher, Harold S.,** *Critique of Radiometric Dating,* Institute for Creation Research, San Diego, 1973, 46 pp.

7. **Moore, John N.** and **Slusher, Harold S.,** Eds., *Biology: A Search for Order in Complexity,* Zondervan, Grand Rapids, MI, 1970, 548 pp.

8. **Morris, Henry M.** and **Gish, Duane T.,** *The Battle for Creation,* Creation-Life Publishers, San Diego, 1976, 321 pp.

*9. **Clark, Marlyn E.,** *Our Amazing Circulatory System,* Creation-Life Publishers, San Diego, 1976, 66 pp.

Programmed Instruction Books

1. **Parker, Gary E.** and **Mertens, Thomas R.,** *Life's Basis: Biomolecules,* John Wiley & Sons, New York, 1973, 158 pp.
2. **Parker, Gary E., Reynolds, W.A.,** and **Reynolds, R.,** *DNA: The Key to Life, 2nd ed.,* Educational Methods, Chicago, 1975, 152 pp.

Two-Model Books

1. **Bliss, Richard B.,** *Origins: Two Models,* Creation-Life Publishers, San Diego, 1978, 56 pp.

BIBLIOGRAPHY

1. **Abelson, P.H.,** *Proceedings of National Academy of Science, 1966, 55:*1365
2. **Davidson, C.F.,** The Precambrian Atmosphere: *Nature,* v. 197, 1963, p. 893.
3. **Austin, S.A.,** *Did the Early Earth Have a Reducing Atmosphere?,* Institute for Creation Research, San Diego, 1982, p. 4.
4. **Miller, S.** and **L. Orgel,** *The Origins of Life on the Earth,* Prentice-Hall, Inc., New Jersey, 1974, pp. 85, 145
5. **Gish, D. T.,** *Speculations and Experiments Related to Theories on the Origin of Life,* Institute for Creation Research, San Diego, 1972, p. 29
6. **Hull, D. E.,** *Nature,* 1960, *186:*694
7. **Fox, S. W.** and **K. Harada,** "Thermal copolymerization of amino acids to a product resembling protein," *Science* 128 (1958):1214
8. **Parker, G. E.,** *Creation Research Society Quarterly,* 1970, *7:*98
9. **Oparin, A. I.,** *Origin of Life on Earth,* Academic Press, Inc., New York, 1965, pp. 133, 176

*Books that are somewhat technical in nature.

10. **O'Connor, R. F.,** *Chemical Principles & Their Biological Implications,* Wiley, 1974, p. 413

11. **Green, D. E.** and **Goldberge, R.F.,** *Molecular Insights into the Living Process,* New York: Academic Press, 1967, p. 407

12. **Salisbury, F. B.,** *Nature,* 1969, *224*:342.

CREDITS

Designed and illustrated by Marvin Ross, Art Director,
Institute for Creation Research

Figure 2.1, Tim Ravenna. **Figure 2.2,** University of
California at San Diego.
Front cover photograph by Steve Pittman

NOTES

NOTES